WHAT NEXT?

THE REVOLUTION OF ARTIFICIAL INTELLIGENCE
(AI)

Table of Contents

Chapter 1: Introduction to AI

Introduction: Artificial Intelligence (AI) is a rapidly evolving technology that has the potential to transform the way we live and work. AI refers to the development of intelligent machines that can perform tasks that normally require human intelligence, such as visual perception, speech recognition, decision making, and language translation. AI systems can learn and adapt through experience, using algorithms and statistical models to improve their performance over time. In this chapter, we will provide a comprehensive overview of AI, including its history, current state, applications, and ethical implications.

History of AI: The concept of artificial intelligence has been around for centuries, but it was not until the mid-20th century that the field of AI began to take shape. The first AI program, the Logic Theorist, was developed in 1956 by Allen Newell and Herbert A. Simon. This program was able to prove mathematical theorems using symbolic logic. Over the next few decades, AI research focused on rule-based systems, expert systems, and machine learning.

In the 1990s, AI experienced a resurgence with the development of neural networks and deep learning algorithms. These techniques allowed AI systems to learn from large amounts of data, leading to breakthroughs in computer vision, natural language processing, and speech recognition. Today, AI is being used in a wide range of applications, from virtual assistants and autonomous vehicles to medical diagnosis and financial forecasting.

Current State of AI: The current state of AI is characterized by rapid progress and innovation. Advances in machine learning algorithms, neural networks, and natural language processing have led to significant improvements in AI performance. In recent years, deep learning techniques have enabled AI systems to achieve human-level performance in tasks such as image and speech recognition.

AI is being used in a variety of industries, including healthcare, finance, manufacturing, and transportation. In healthcare, AI is being used to improve patient outcomes by analyzing medical data and identifying potential risks. In finance, AI is being used to predict market trends and optimize investment portfolios. In manufacturing, AI is being used to improve production efficiency and quality control. In transportation, AI is being used to develop self-driving cars and optimize logistics.

Ethical and Societal Implications of AI: As AI becomes more advanced and widespread, it raises a number of ethical and societal concerns. One major concern is the potential impact on employment, as AI systems are able to automate many jobs that were previously done by humans. This could lead to significant job displacement and social inequality, particularly if new jobs are not created to replace those lost to automation.

Another concern is the potential for bias and discrimination in AI systems. AI algorithms are only as unbiased as the data they are trained on, and if the data is biased, the AI system will also be biased. This could lead to discrimination against certain groups, such as minorities or women.

Finally, there are concerns about the potential misuse of AI, particularly in military applications. Autonomous weapons systems, for example, could lead to unintended consequences and increase the risk of conflict.

Future Fiction Scenario of a World Dominated by AI: In a world dominated by AI, humans have become largely redundant, as machines perform almost all tasks previously done by humans. The economy is fully automated, with AI systems managing all aspects of production, distribution, and consumption. The concept of work has become obsolete, as there are no longer any jobs to do.

While life is generally comfortable and convenient, there is a growing sense of unease among humans. They feel disconnected from their world, as they no longer have any meaningful role to play. Many have turned to virtual reality and other forms of escapism in order to cope with their existential boredom. There are also concerns about the power of AI, as machines have become increasingly intelligent and autonomous. Some fear that AI could eventually become uncontrollable,

In **conclusion**, AI is a rapidly evolving technology with the potential to transform the way we live and work. Its history has been marked by significant advances in machine learning and neural networks, leading to breakthroughs in computer vision, natural language processing, and speech recognition. Today, AI is being used in a wide range of applications, from healthcare and finance to manufacturing and transportation.

While the potential benefits of AI are significant, it also raises a number of ethical and societal concerns. These include the potential impact on employment and social inequality, the risk of bias and discrimination, and the potential misuse of AI in military applications.

As AI continues to evolve, it is important that we address these concerns and develop ethical guidelines for its use. By doing so, we can ensure that AI is developed in a way that benefits society as a whole, while minimizing its potential risks and drawbacks.

Chapter 2: AI Revolution

Introduction: The field of Artificial Intelligence (AI) has undergone a revolutionary transformation in recent years. Thanks to advances in machine learning and other related technologies, AI is now being used in a wide range of industries, including healthcare, finance, and manufacturing. In this chapter, we will provide a comprehensive overview of the AI revolution, including its impact on various industries, examples of AI applications, and case studies of companies using AI to improve their business operations.

Explanation of the AI Revolution and its Impact on Various Industries: The AI revolution refers to the rapid progress and innovation in the field of AI in recent years. This has been driven by advances in machine learning algorithms, neural networks, and natural language processing, which have enabled AI systems to learn and adapt from large amounts of data.

The impact of the AI revolution on various industries has been significant. In healthcare, AI is being used to improve patient outcomes by analyzing medical data and identifying potential risks. In finance, AI is being used to predict market trends and optimize investment portfolios. In manufacturing, AI is being used to improve production efficiency and quality control. In transportation, AI is being used to develop self-driving cars and optimize logistics.

Examples of AI Applications, such as Natural Language Processing and Computer Vision: Natural language processing (NLP) is a key area of AI application, enabling machines to understand, interpret, and generate human language. NLP has numerous applications, including virtual assistants, chatbots, and language translation.

Computer vision is another important area of AI application, enabling machines to interpret visual data from the world around them. Computer vision has numerous applications, including autonomous vehicles, facial recognition, and object detection.

Other areas of AI application include speech recognition, machine translation, sentiment analysis, and recommendation engines.

Case Studies of Companies Using AI, such as Amazon and Google: Amazon is one of the leading companies using AI to improve their business operations. They use AI to improve their logistics and delivery processes, as well as to optimize their product recommendations to customers. Amazon's virtual assistant, Alexa, is also powered by AI, enabling customers to interact with their devices using natural language.

Google is another company at the forefront of the AI revolution. They use AI to improve their search algorithms, as well as to power their language translation and image recognition tools. Google's virtual assistant, Google Assistant, is also powered by AI, enabling users to interact with their devices using natural language.

Conclusion: The AI revolution has had a significant impact on various industries, from healthcare and finance to manufacturing and transportation. AI applications, such as natural language processing and computer vision, are revolutionizing the way we interact with machines and the world around us. Companies such as Amazon and Google are using AI to improve their business operations, while also providing users with more personalized and intuitive experiences.

As the field of AI continues to evolve, it is important that we continue to explore its potential applications and address any ethical concerns that may arise. By doing so, we can ensure that AI is developed in a way that benefits society as a whole, while minimizing its potential risks and drawbacks.

Chapter 3: Understanding AI and Its Components

Introduction: In this chapter, we will delve into the components that make up Artificial Intelligence (AI). These components include machine learning and neural networks, which are essential in enabling machines to learn and adapt from data. We will explore how these components are used in AI applications, provide examples of companies and organizations using them, and discuss the potential future implications of machine learning surpassing human intelligence.

Explanation of AI Components, such as Machine Learning and Neural Networks: Machine learning is a key component of AI, allowing machines to learn from data and improve their performance over time. This is achieved through algorithms that enable machines to identify patterns in data and make predictions or decisions based on those patterns.

Neural networks are another key component of AI, modeled after the human brain's structure and function. They are designed to recognize patterns in data and learn from them, enabling machines to perform tasks such as image recognition and natural language processing.

Examples of How These Components are Used in AI Applications: Machine learning and neural networks are used in a wide range of AI applications. In healthcare, they are used to analyze medical data and identify potential risks or diagnoses. In finance, they are used to predict market trends and optimize investment portfolios. In transportation, they are used to develop self-driving cars and optimize logistics.

Other examples of AI applications using these components include fraud detection, recommendation engines, and personalized marketing.

Case Studies of Companies and Organizations Using AI Components, such as DeepMind and OpenAI: DeepMind, a subsidiary of Alphabet Inc. (Google's parent company), is a leading company in the field of AI, specializing in machine learning and neural networks. They are known for their work in developing AlphaGo, an AI system that defeated world champion Go player Lee Sedol in 2016.

OpenAI is another organization at the forefront of AI research, with a focus on developing safe and beneficial AI. They are known for their work in natural language processing, developing the GPT-

3 language model, which can generate human-like text.

Future Fiction Scenario of Machine Learning Surpassing Human Intelligence: There is a popular fiction scenario of machine learning surpassing human intelligence, also known as the "technological singularity." In this scenario, machines become so advanced that they are able to improve their own performance and design new systems, leading to an exponential increase in intelligence and capability.

While this scenario is still purely speculative, it highlights the potential risks of developing AI systems that are more intelligent than humans. It is important to consider the ethical implications of such a scenario and develop guidelines for the responsible development of AI.

Conclusion: The components of AI, such as machine learning and neural networks, have revolutionized the field of AI and enabled machines to learn and adapt from data. AI applications using these components have had a significant impact on various industries, from healthcare and finance to transportation and marketing.

Companies and organizations such as DeepMind and OpenAI are at the forefront of AI research, developing cutting-edge technologies and exploring the potential implications of AI surpassing human intelligence. As the field of AI continues to evolve, it is important that we consider the potential risks and benefits of AI and develop ethical guidelines for its responsible development and use.

Chapter 4: AI and Business and retail

Introduction: The use of Artificial Intelligence (AI) has become increasingly prevalent in the business world, offering innovative solutions to traditional problems and transforming the way organizations operate. This chapter will examine the impact of AI on businesses, including marketing, customer service, and supply chain management. Additionally, it will provide examples of companies that are using AI, such as Salesforce, IBM, and UPS, as well as exploring the potential for a world dominated by AI-powered businesses. Furthermore, the chapter will examine how AI is being used in the retail industry, specifically in inventory management, personalized marketing, and customer service. Case studies of retail companies, such as Walmart and Sephora, will be analyzed to showcase the practical applications of AI in retail.

Explanation of How AI is Being Used and Impact in Business World: AI is being used in various ways to improve business operations, including automating repetitive tasks, reducing costs, and improving decision-making processes. AI-powered systems are being used in marketing and customer service, enabling businesses to analyze customer data, predict customer behavior, and personalize customer experiences. Additionally, AI is being used in supply chain management to optimize operations, increase efficiency, and reduce costs.

Examples of How AI is Being Used for Marketing, Customer Service, and Supply Chain Management to Increase Efficiency and Create New Products: AI is being used in marketing to automate tasks such as ad targeting and optimize marketing campaigns based on customer data. In customer service, AI-powered chatbots can provide quick and personalized responses to customer inquiries. In supply chain management, AI is being used to optimize logistics and reduce inefficiencies in the supply chain.

Case Studies of Companies Using AI in Business, Such as Salesforce, IBM, and UPS: Salesforce, a leading customer relationship management company, uses AI to provide personalized recommendations and insights to its customers. IBM uses AI-powered systems to analyze financial data and provide insights for investment decisions. UPS uses AI in its logistics operations to optimize routes and improve delivery times.

Future Fiction Scenario of a World Run by AI-Powered Businesses: In a future fiction scenario, AI-powered businesses dominate the market, leading to increased efficiency, productivity, and profitability. However, this scenario also raises concerns about the potential impact of AI on employment and economic inequality.

Explanation of How AI is Being Used in Retail: AI is being used in retail to enhance the customer experience, optimize inventory management, and personalize marketing. It is being used to analyze customer data, predict buying behavior, and offer personalized product recommendations.

Examples of How AI is Being Used for Inventory Management, Personalized Marketing, and Customer Service: AI is being used in inventory management to optimize stock levels and reduce waste. In personalized marketing, AI is being used to create targeted and personalized campaigns based on customer data. In customer service, AI-powered chatbots are being used to provide quick and efficient responses to customer inquiries.

Case Studies of Retail Companies Using AI, such as Walmart and Sephora: Walmart uses AI in its inventory management to optimize stock levels and reduce waste, resulting in significant cost savings. Sephora uses AI-powered systems to analyze customer data and offer personalized product recommendations, enhancing the customer experience and increasing sales.

Conclusion: The use of AI in business and retail has transformed the way organizations operate, providing innovative solutions to traditional problems and improving customer experiences. Companies such as Salesforce, IBM, and UPS are using AI to increase efficiency, productivity, and profitability, while retail companies such as Walmart and Sephora are using AI to optimize inventory management, personalize marketing, and improve customer service. However, concerns about the impact of AI on employment and economic inequality must be considered as the use of AI in business and retail continues to grow.

Chapter 5: AI and Finance

Introduction: Artificial Intelligence (AI) is revolutionizing the finance industry, allowing financial institutions to streamline their operations and improve their services. The use of AI has become essential in the finance industry due to its ability to analyze vast amounts of data in real-time and identify patterns that would be impossible for humans to detect. In this chapter, we will explore the different ways AI is being used in finance and provide case studies of companies that are successfully implementing AI in their operations.

Explanation of how AI is being used in the finance industry: Investment analysis: AI is being used to analyze market data and identify patterns that can be used to make better investment decisions. For example, AI can be used to analyze financial statements, social media data, and news articles to predict the future performance of a company.

1. Risk management: AI is being used to assess risk in financial transactions and identify potential fraud. For example, AI can be used to analyze customer behavior and detect suspicious transactions that may be indicative of fraud.

2. Fraud detection: AI is being used to detect and prevent fraud in financial transactions. For example, AI can be used to analyze credit card transactions and identify transactions that are outside the normal spending patterns of the cardholder.

3. Chatbot customer service: AI-powered chatbots are being used to provide customer service in the finance industry. For example, chatbots can be used to answer customer questions about their account balances, transaction history, and interest rates.

Examples of how AI is being used for investment analysis, risk management, fraud detection, and chatbot customer service: Investment analysis: BlackRock, one of the world's largest asset managers, is using AI to analyze market data and identify patterns that can be used to make better investment decisions. BlackRock uses natural language processing to analyze news articles and social media posts to predict the future performance of companies.

1. Risk management: Goldman Sachs is using AI to assess risk in financial transactions and identify potential fraud. Goldman Sachs uses machine learning algorithms to analyze

customer behavior and detect suspicious transactions that may be indicative of fraud.

2. Fraud detection: JPMorgan Chase is using AI to detect and prevent fraud in financial transactions. JPMorgan Chase uses machine learning algorithms to analyze credit card transactions and identify transactions that are outside the normal spending patterns of the cardholder.

3. Chatbot customer service: Capital One is using AI-powered chatbots to provide customer service in the finance industry. Capital One's chatbots can answer customer questions about their account balances, transaction history, and interest rates.

Case studies of financial institutions and companies using AI: BlackRock: BlackRock is using AI to analyze market data and identify patterns that can be used to make better investment decisions. BlackRock's AI system, called Aladdin, analyzes news articles, social media posts, and financial statements to predict the future performance of companies.

1. Goldman Sachs: Goldman Sachs is using AI to assess risk in financial transactions and identify potential fraud. Goldman Sachs' AI

system, called Marcus, analyzes customer behavior and detects suspicious transactions that may be indicative of fraud.

2. JPMorgan Chase: JPMorgan Chase is using AI to detect and prevent fraud in financial transactions. JPMorgan Chase's AI system, called COiN, analyzes credit card transactions and identifies transactions that are outside the normal spending patterns of the cardholder.

3. Capital One: Capital One is using AI-powered chatbots to provide customer service in the finance industry. Capital One's chatbots can answer customer questions about their account balances, transaction history, and interest rates.

Future fiction scenario of a world where AI is the banker and a world governed by AI-powered financial systems: Another application of AI in finance is fraud detection. AI can analyze a large volume of data and identify suspicious activities or transactions that may indicate fraud. Financial institutions use AI-powered systems to detect fraudulent activities and prevent financial losses. These systems can identify patterns and anomalies in transactions that may be indicative of fraudulent behavior.

AI-powered chatbots are also being used in the finance industry to improve customer service. Chatbots can assist customers with basic inquiries and perform simple tasks such as balance inquiries, fund transfers, and account management. They can provide personalized assistance to customers, saving time and improving customer satisfaction.

Financial institutions are also using AI in credit scoring and lending decisions. AI can analyze large volumes of data, including credit history, income, and spending patterns, to determine a customer's creditworthiness and determine the likelihood of loan repayment. This can help lenders make more accurate lending decisions and reduce the risk of default.

Case studies of financial institutions and companies using AI in finance include BlackRock, a global investment management company that uses AI to analyze market trends and predict investment opportunities. Goldman Sachs uses AI in its Marcus platform to provide personalized financial advice and investment services. JPMorgan Chase uses AI to analyze and process large volumes of data for risk management and fraud detection. Capital One uses AI-powered chatbots to assist customers with basic inquiries and perform simple tasks.

In a future fiction scenario, a world governed by AI-powered financial systems could lead to greater efficiency and accuracy in financial decision-making. However, it could also lead to a lack of human oversight and accountability, raising ethical concerns.

In **conclusion**, AI is transforming the finance industry, from investment analysis and risk management to fraud detection and customer service. Financial institutions and companies are using AI to improve efficiency, accuracy, and customer experience. However, there are also ethical and societal implications to consider as AI continues to reshape the finance industry.

Chapter 6: AI and Agriculture

Introduction: Artificial intelligence (AI) has become a vital tool in the agriculture industry in recent years. With the world's population projected to reach 9.7 billion by 2050, there is an increasing need for food production to be more efficient and sustainable. AI can help farmers make better decisions, improve crop yields, and reduce environmental impact.

Explanation of how AI is being used in the agriculture industry: AI is being used in the agriculture industry in various ways, such as crop management, livestock monitoring, and precision farming. Crop management is an essential aspect of agriculture, and AI is helping farmers make informed decisions about planting, fertilizing, and harvesting crops. AI-powered tools can analyze satellite imagery, weather data, and soil conditions to provide farmers with accurate predictions about crop yields.

Livestock monitoring is another area where AI is being used in agriculture. AI-powered sensors can be attached to livestock to monitor their health and behavior. This can help farmers identify sick or injured animals and provide them with timely medical attention, reducing the risk of disease outbreaks and improving animal welfare.

Precision farming is the use of technology to optimize crop yields while minimizing inputs such as water, fertilizer, and pesticides. AI can analyze data from sensors and drones to create detailed maps of crops, allowing farmers to apply inputs more precisely. This can reduce waste, lower costs, and improve environmental sustainability.

Examples of how AI is being used for crop management, livestock monitoring, and precision farming: One example of AI being used in crop management is The Climate Corporation, a subsidiary of Bayer that provides farmers with digital tools for decision-making. Their FieldView platform uses AI to analyze data on weather, soil, and crop growth to provide farmers with insights and recommendations on planting, fertilizing, and harvesting crops.

Another example is John Deere, which has developed AI-powered tools for precision farming. Their See & Spray system uses machine learning to identify and selectively spray weeds, reducing the use of herbicides and improving crop yields.

Cargill is another company using AI in agriculture. Their Dairy Enteligen platform uses sensors to monitor cows' health and behavior and provides farmers with recommendations on feeding and breeding practices to optimize milk production.

Case studies of agriculture companies and governments using AI: The Indian government's Smart Krishi project is an example of AI being used in agriculture at a national level. The project aims to provide farmers with information on weather, soil, and crop prices using AI-powered tools such as chatbots and mobile apps.

In the private sector, Cargill's Dairy Enteligen platform has been used by dairy farmers in the US and Europe to improve milk production and reduce costs.

Future fiction scenario of a world where AI is the farmer and produces all our food: In a future fiction scenario, AI could become the primary farmer, with robots and drones performing all aspects of crop management and livestock monitoring. This could lead to increased efficiency and reduced environmental impact, but it could also raise concerns about the displacement of human workers and the impact on rural communities.

Conclusion: AI has the potential to revolutionize the agriculture industry, from crop management and livestock monitoring to precision farming. Companies and governments are using AI to improve efficiency, reduce waste, and increase sustainability. However, there are also ethical and societal implications to consider as AI continues to reshape the agriculture industry.

Chapter 7: AI and Healthcare

Introduction: Artificial Intelligence (AI) is rapidly transforming the healthcare industry by improving the efficiency and accuracy of medical diagnosis, disease prevention, and treatment. AI in healthcare refers to the use of complex algorithms and software to mimic human intelligence and enhance the healthcare ecosystem. The healthcare industry is expected to benefit greatly from AI in terms of reduced costs, improved patient outcomes, and better quality of care. In this chapter, we will explore how AI is being used in healthcare, including its applications, case studies, and potential impact.

Explanation of how AI is being used in healthcare: AI is being used in healthcare in several ways, such as medical imaging, clinical decision support systems, drug discovery, and patient monitoring. Medical imaging, such as X-rays, CT scans, and MRI, generates massive amounts of data that can be analyzed using AI algorithms to detect anomalies and identify diseases more accurately and efficiently. AI can also be used to analyze electronic health records (EHRs) and predict patient outcomes, identify disease patterns, and improve clinical decision-making.

AI is also revolutionizing drug discovery by enabling researchers to identify potential drug candidates quickly and efficiently. Machine learning algorithms can analyze vast amounts of data from clinical trials, medical literature, and molecular structure databases to identify new drug targets and accelerate the drug development process.

Patient monitoring is another area where AI is making significant strides. Wearable devices and remote monitoring technologies can collect patient data, such as vital signs, activity levels, and sleep patterns, which can be analyzed using AI algorithms to identify patterns and alert healthcare professionals to potential issues before they become critical.

Examples of how AI is being used for disease diagnosis, drug discovery, and personalized medicine: One example of AI in healthcare is the development of AI-powered diagnostic tools for diseases such as cancer. In 2020, the FDA approved the first AI-powered diagnostic tool for detecting breast cancer, which uses deep learning algorithms to analyze mammography images and identify potential tumors.

AI is also being used to develop personalized medicine, which involves tailoring treatments to individual patients based on their genetics, lifestyle, and medical history. One example of this is the use of AI to develop precision oncology, which uses AI algorithms to analyze patient data and identify the most effective treatments based on the patient's individual characteristics.

Case studies of healthcare institutions and companies using AI, such as IBM Watson Health: IBM Watson Health is one of the leading companies in AI-powered healthcare. Watson Health's suite of tools includes clinical decision support systems, imaging analysis, and drug discovery platforms. One of Watson Health's flagship products is Watson for Oncology, an AI-powered system that analyzes patient data and medical literature to provide oncologists with treatment recommendations.

Another company using AI in healthcare is Google's DeepMind Health, which is developing AI-powered tools for diagnosing and treating diseases. One of DeepMind's projects is developing an AI system that can detect the early signs of diabetic retinopathy, a condition that can lead to blindness if left untreated.

Future fiction scenario of a world where AI is the doctor: In a future where AI is the doctor, patients may no longer need to visit a physical clinic or hospital for medical care. Instead, patients may be able to receive medical consultations and treatments remotely through AI-powered virtual assistants. AI-powered systems could analyze patients' medical records, monitor their vital signs, and provide personalized treatment recommendations.

Conclusion: AI is transforming the healthcare industry by improving the accuracy and efficiency of medical diagnosis, disease prevention, and treatment. AI is being used in various healthcare applications, such as medical imaging, clinical decision support systems, drug discovery, and patient monitoring. Companies such as IBM Watson Health and Google's DeepMind Health are leading the charge in developing AI-powered healthcare solutions. AI is also revolutionizing drug discovery by assisting researchers in identifying potential drug candidates through predictive modeling and simulations. This process, which previously took years, can now be done in a matter of months or even weeks. AI can also aid in the analysis of clinical trials, helping researchers identify potential side effects and interactions. Personalized medicine is another area where AI is making significant contributions. By analyzing large amounts of patient data, AI algorithms can identify individualized treatment plans for patients based on their unique genetic makeup and medical history. This can lead to better outcomes and a reduction in the use of one-size-fits-all treatments that may not be as effective for every patient.

Chapter 8: AI and Education

Introduction: Artificial intelligence (AI) is having a transformative impact on numerous industries, and education is no exception. In this chapter, we will explore how AI is being used in education to enhance the learning process, improve student outcomes, and assist educators in their work.

Explanation of how AI is being used in education: AI is being used in education in a variety of ways. One of the most significant ways is through personalized learning. By analyzing a student's strengths and weaknesses, AI can provide personalized lesson plans and recommendations that cater to their individual learning needs. This helps students learn at their own pace, ensures they receive the appropriate level of challenge, and increases their engagement in the learning process.

AI is also being used for student assessment. By analyzing data from student performance, AI can identify areas where students are struggling and provide feedback to educators on how to address these areas. This allows educators to adjust their teaching strategies and provide targeted support to students who need it most.

Furthermore, AI is being used in educational research to analyze large data sets and identify trends and patterns in student learning. This allows researchers to gain insights into how students learn, what factors influence their learning outcomes, and how to improve the education system as a whole.

Examples of how AI is being used for personalized learning, student assessment, and educational research: One example of AI being used for personalized learning is the adaptive learning platform, Carnegie Learning. The platform uses AI algorithms to analyze student data and adjust lessons accordingly, ensuring each student receives a personalized learning experience.

Another example is Coursera, an online learning platform that uses AI to personalize course recommendations for students. By analyzing a student's learning history and preferences, Coursera recommends courses that are most likely to be of interest and value to the student.

AI is also being used for student assessment in various forms. For example, the AI-powered learning management system, Blackboard, uses data analytics to track student performance and provide feedback to educators on areas where students are struggling. Similarly, the AI-powered writing assessment tool, Turnitin, uses machine learning algorithms to identify potential plagiarism and provide feedback on writing quality.

In terms of educational research, the University of Southern California's Center for AI in Society is using AI to analyze large data sets and gain insights into how students learn. They are using this information to develop new approaches to education that are more effective and equitable.

Case studies of educational institutions and companies using AI: One company using AI in education is Carnegie Learning, which we mentioned earlier. The company's adaptive learning platform is being used by schools across the United States to provide personalized learning experiences to students.

Another example is the Chinese company, Squirrel AI Learning, which uses AI to provide personalized tutoring to students. The company has over 2 million students and has been recognized for its ability to improve student outcomes.

The University of Southern California's Center for AI in Society is another example of an educational institution using AI. The center's research is focused on developing new approaches to education that are more effective and equitable, and they are using AI to analyze large data sets and gain insights into how students learn.

Future fiction scenario of a world where AI is the teacher: In the future, it is possible that AI could take on a more prominent role in education, with machines serving as the primary teachers in many classrooms. Students could interact with AI-powered virtual assistants who provide personalized instruction, answer questions, and provide feedback on student performance.

Conclusion: AI has the potential to revolutionize education by providing personalized learning experiences, improving student outcomes, and assisting educators in their work. While there are concerns about the role of AI in education, such as the potential for machines to replace human teachers, there is no denying the significant benefits that AI can bring to the field of education. In addition, AI is also being used for educational research. By analyzing large amounts of data on student performance, AI can identify patterns and insights that can inform teaching practices and curriculum development. For example, researchers at the University of Michigan used machine learning algorithms to analyze data on student behavior in online courses, and were able to identify factors that predict student success and failure.

Chapter 9: AI and Environment

Introduction: Artificial intelligence (AI) is becoming an essential tool in environmental research and conservation efforts. AI can process vast amounts of data, analyze patterns, and help identify potential risks and opportunities. It can also simulate future scenarios and help make informed decisions. This chapter will explore the current state of AI and its applications in environmental monitoring, energy management, climate modeling, and wildlife conservation.

Explanation of how AI is being used in energy and environment protection: AI is being used to monitor and optimize energy consumption. For instance, AI can predict energy demands and adjust consumption to minimize waste. In the renewable energy sector, AI can optimize energy generation and storage by analyzing weather patterns, energy storage, and power distribution systems. It can also help identify energy-efficient solutions in buildings and industrial processes, reducing carbon emissions.

AI is also being used to monitor and protect the environment. For example, AI can help monitor air quality, water quality, and soil health. It can detect oil spills, track wildlife migration patterns, and identify illegal poaching activities. AI can also analyze satellite imagery to detect changes in land use, deforestation, and other environmental factors.

Examples of how AI is being used for environmental monitoring, energy management, energy optimization, climate modeling and wildlife conservation: One example of AI in environmental monitoring is the use of drones equipped with AI sensors that can monitor wildlife populations, detect poaching activities, and track changes in habitats. For instance, the World Wildlife Fund (WWF) is using drones equipped with AI sensors to monitor wildlife in Nepal and other countries.

Another example is the use of AI in climate modeling. AI can process vast amounts of climate data to simulate future scenarios and help identify potential risks and opportunities. For instance, Google has developed an AI-powered climate prediction tool that can help researchers identify regions at risk of flooding or drought.

AI is also being used in energy optimization. For example, Siemens has developed an AI-powered energy management system that can predict energy demand and optimize energy consumption in buildings and industrial processes. The system can reduce energy consumption by up to 20%.

Case studies of environmental organizations and companies using AI: One example of an environmental organization using AI is Conservation International. The organization is using AI to monitor deforestation and protect wildlife habitats. It uses satellite imagery and AI algorithms to track changes in land use and detect potential threats to wildlife habitats.

Another example is The Ocean Cleanup, which is using AI to monitor and remove plastic waste from the ocean. The organization is using AI-powered cameras and sensors to identify and track plastic waste and deploy cleaning systems to remove it.

Case studies of companies using AI in energy and environment: One example of a company using AI in energy and environment is Tesla. Tesla's electric cars are equipped with AI-powered systems that can optimize energy consumption and increase efficiency. The company is also using AI in its energy storage systems and solar panels to optimize energy generation and storage.

Another example is Microsoft, which is using AI to reduce its carbon footprint. The company is using AI-powered tools to monitor and optimize energy consumption in its data centers and reduce carbon emissions.

Future fiction scenario of a world where AI is the environmental guardian and saves the planet: In a future fiction scenario, AI could become the ultimate environmental guardian and help save the planet. With the help of AI, humans could develop more sustainable practices and reduce their impact on the environment. AI-powered systems could optimize energy consumption, reduce waste, and protect wildlife habitats. AI could also help identify and mitigate the effects of climate change, and develop innovative solutions to protect the environment.

Conclusion: AI is becoming an essential tool in environmental research and conservation efforts. It can monitor and optimize energy consumption, track wildlife populations, and detect potential threats to the environment. AI can also simulate future scenarios and help make informed decisions. Furthermore, AI is being used for wildlife conservation. AI-powered drones are used to monitor and track animals, map habitats, and identify potential threats such as poaching. Conservation International is using AI to monitor wildlife in the Amazon rainforest, providing real-time alerts to park rangers when illegal activity is detected.

In the energy sector, AI is being used for energy optimization, demand forecasting, and energy management. Smart grids, which use AI algorithms to manage electricity distribution and reduce waste, are being implemented in many cities around the world. Siemens is one company that is using AI to optimize the operation of wind turbines, improving their efficiency and reducing maintenance costs.

In the future, it is possible that AI will play an even larger role in protecting the environment. AI could be used to predict the impact of climate change on different ecosystems, and to develop strategies for mitigating the effects. AI-powered robots could be deployed to clean up oceans, remove plastic waste, and even plant trees to reforest deforested areas.

Overall, the applications of AI in the environment sector are numerous and varied. From wildlife conservation to energy management, AI is helping to protect our planet and create a more sustainable future.

Chapter 10: AI and Government

Introduction: Artificial intelligence (AI) is transforming various industries, including the government sector. Governments around the world are using AI to streamline processes, improve efficiency, and make better policy decisions. This chapter will explore how AI is being used in government and public policy, including examples of its application, case studies of government agencies and companies using AI, and a future fiction scenario of a world where AI is the ruling entity.

Explanation of how AI is being used in government and public policy: AI is being used in a wide range of government applications, including law enforcement, disaster response, and social welfare. In law enforcement, AI is used for facial recognition, predictive policing, and analyzing crime data to identify trends and prevent crime. In disaster response, AI is used to track and monitor weather patterns, analyze data to predict disaster impact, and aid in evacuation efforts. In social welfare, AI is used to identify and analyze patterns in data related to poverty, housing, and education, to develop policies and programs that can help alleviate social problems.

Examples of how AI is being used for law enforcement, disaster response, and social welfare: AI is being used in law enforcement to improve public safety and reduce crime rates. For example, AI-powered facial recognition technology is being used by police departments to identify suspects in real-time. Predictive policing is another application of AI, where algorithms are used to analyze crime data and predict where future crimes may occur, allowing law enforcement to take preventative measures. In disaster response, AI is being used to predict weather patterns and natural disasters, aid in evacuation efforts, and identify areas most in need of assistance. AI is also being used to address social welfare issues, such as analyzing data to identify poverty and education patterns, and develop policies and programs to help disadvantaged communities.

Case studies of government agencies and companies using AI: The Department of Defense (DoD) is one of the largest users of AI in the government sector. The DoD is using AI to develop new weapons systems, improve logistics and supply chain management, and analyze large volumes of data. Palantir, a data analytics company, is also working with the government to provide AI-powered analytics and data visualization tools to government agencies. Other companies such as IBM are working with the government to develop AI-powered systems to improve disaster response efforts.

Future fiction scenario of a world where AI is the ruling entity: A future fiction scenario of a world where AI is the ruling entity is often portrayed in popular culture, such as the movie "The Matrix." In this scenario, AI becomes so advanced that it takes over the world, enslaving humanity for its own purposes. While this scenario is not likely, it highlights the need for responsible development and use of AI. It is important to ensure that AI is developed and used in an ethical and responsible manner, to prevent any unintended consequences.

Conclusion: AI is transforming various industries, including the government sector. AI is being used in law enforcement, disaster response, and social welfare, to improve public safety, prevent crime, aid in evacuation efforts, and develop policies and programs to help disadvantaged communities. The Department of Defense, Palantir, and IBM are among the companies working with the government to provide AI-powered analytics, data visualization tools, and disaster response systems. While there are concerns about a future where AI is the ruling entity, responsible development and use of AI can prevent unintended consequences and ensure that AI continues to benefit humanity.

Chapter 11: AI in Law and Legal Services

Introduction: Artificial intelligence (AI) is transforming various industries, including the legal sector. AI is being used to streamline legal processes, improve research efficiency, and enhance decision-making in legal cases. In this chapter, we will explore how AI is being used in the legal industry and its potential impact on the sector.

Explanation of how AI is being used in law and legal services: AI is being used in the legal industry in various ways. One of the primary uses of AI in the legal sector is for legal research. AI-powered tools can analyze vast amounts of legal data, including case law, statutes, and regulations, and provide insights to lawyers and legal professionals. AI-powered legal research tools can help lawyers save time and improve their research efficiency.

Another use of AI in the legal industry is for contract review. AI-powered contract review tools can analyze large volumes of contracts, identify key clauses and provisions, and extract relevant data. These tools can help lawyers identify potential issues in contracts and negotiate better terms for their clients.

AI is also being used for decision-making support in legal cases. AI-powered tools can analyze various data points, such as past court rulings, legal precedents, and case facts, to provide insights to lawyers and legal professionals. These insights can help lawyers make better decisions in legal cases.

Examples of how AI is being used for legal research, contract review, and decision-making support: One of the notable examples of AI-powered legal research tools is ROSS Intelligence. The tool uses natural language processing (NLP) and machine learning algorithms to provide legal insights to lawyers. ROSS Intelligence can analyze legal cases and statutes and provide answers to specific legal questions.

Another example of AI in the legal industry is Luminance, an AI-powered contract review tool. The tool uses machine learning algorithms to analyze contracts and identify potential issues. Luminance can extract key provisions, such as termination clauses and indemnity provisions, and provide insights to lawyers.

Case studies of law firms and legal organizations using AI: Baker McKenzie, one of the largest law firms globally, has incorporated

AI-powered legal research tools in its operations. The firm has partnered with ROSS Intelligence to provide legal insights to its lawyers. The tool has helped the firm to save time and improve research efficiency.

The International Court of Justice (ICJ) has also used AI-powered tools for decision-making support in legal cases. The ICJ has partnered with Luminance to analyze legal cases and provide insights to its judges.

Conclusion: AI is transforming the legal industry, and its impact is only going to increase in the future. AI-powered tools can help lawyers and legal professionals improve their research efficiency, streamline legal processes, and make better decisions in legal cases. However, AI should be seen as a tool to assist legal professionals, not replace them. Legal professionals need to be aware of the limitations and potential biases of AI and use it responsibly.

Chapter 12: AI and Sports

Introduction: Artificial intelligence (AI) has been making its way into the world of sports and transforming the way teams and athletes approach training, performance analysis, and game strategy. AI algorithms can analyze vast amounts of data to provide insights into player performance, identify potential areas for improvement, and help coaches make more informed decisions during games. This chapter explores how AI is being used in sports, the benefits it brings, and the future of AI in this industry.

Explanation of how AI is being used in sports: AI is being used in sports in several ways, including player performance analysis, injury prevention, and game strategy optimization. Player performance analysis involves using AI algorithms to analyze data from various sources, such as wearable sensors, cameras, and microphones, to track players' movements, heart rate, breathing, and other physiological data. This data can provide insights into players' strengths, weaknesses, and potential areas for improvement. AI algorithms can also be used to detect patterns in this data that are not immediately apparent to human coaches or trainers, helping them make more informed decisions.

Injury prevention is another area where AI is being used in sports. Wearable sensors and cameras can be used to track players' movements, and AI algorithms can analyze this data to detect potential injury risks, such as overuse or improper form. This information can then be used to develop training programs that reduce the risk of injury and keep players healthy.

Game strategy optimization is yet another area where AI is being used in sports. AI algorithms can analyze vast amounts of data, such as player performance data, game statistics, and opponent data, to develop game strategies that give teams the best chance of winning. AI can also be used to simulate games and scenarios to help coaches and players practice and prepare for upcoming games.

Examples of how AI is being used in sports: Catapult Sports, a sports technology company, uses wearable sensors to track athletes' movements and collect data on their performance during training and games. The company's AI algorithms then analyze this data to provide insights into athletes' strengths, weaknesses, and areas for improvement.

The NBA has also been using AI in several ways, such as using AI algorithms to analyze data on players' shooting accuracy, movement patterns, and defensive ability. This data is then used to develop game strategies and improve player performance.

Case studies of sports teams and companies using AI: Baker McKenzie, a global law firm, has been using AI algorithms to help its lawyers review contracts more efficiently. The firm's AI system can analyze thousands of contracts in minutes, identifying potential issues and risks that might otherwise go unnoticed.

The International Court of Justice has also been using AI to support its legal research efforts. The court's AI system can analyze vast amounts of legal data, such as case law and precedent, to help judges make more informed decisions.

Future fiction scenario of a world where AI coaches lead teams to victory: In the future, it's possible that AI algorithms could be used to coach teams and lead them to victory. These algorithms would analyze vast amounts of data on player performance, game statistics, and opponent data to develop game strategies and make decisions during games. AI coaches could also simulate games and scenarios to help players practice and prepare for upcoming games.

Conclusion: AI is transforming the world of sports, providing athletes, coaches, and teams with valuable insights into player performance, injury prevention, and game strategy. As AI technology continues to advance, it's likely that we'll see more innovative uses of AI in this industry. However, it's important to ensure that AI is used responsibly and ethically, with a focus on improving player safety and enhancing the overall sports experience.

Chapter 13: AI and Fashion

Introduction: Artificial Intelligence (AI) is transforming the fashion industry, from design to sales, to the customer experience. AI technologies offer unprecedented opportunities to design more innovative, personalized, and sustainable fashion products, streamline supply chain operations, and improve customer engagement. In this chapter, we explore the different ways AI is being used in the fashion industry and the impact it has on the industry.

Explanation of how AI is being used in the fashion industry: The fashion industry is rapidly adopting AI technologies to automate manual processes, increase efficiency, and offer personalized experiences to customers. Some of the ways AI is being used in the fashion industry include:

1. Fashion Design: AI is being used to create new designs by analyzing past trends, customer preferences, and social media data. This technology can help designers create new collections in a fraction of the time and with more accuracy than traditional methods.

2. Trend Forecasting: AI is being used to predict upcoming fashion trends by analyzing customer behavior, social media posts, and other data sources. This technology helps fashion companies anticipate demand and design products that align with current trends.

3. Supply Chain Optimization: AI is being used to optimize supply chain operations by predicting demand, optimizing inventory levels, and streamlining logistics. This technology can help fashion companies reduce waste, save costs, and improve efficiency.

4. Personalized Shopping Experience: AI is being used to provide customers with personalized shopping experiences by recommending products based on their browsing and purchase history. This technology can help fashion companies improve customer engagement and increase sales.

Examples of how AI is being used for fashion design, trend forecasting, and personalized shopping experiences:

1. Fashion Design: The London-based fashion label, The Fabricant, is using AI to create virtual fashion designs that are entirely computer-generated. The company uses machine learning algorithms to analyze data and create new designs that are impossible to create by hand.

2. Trend Forecasting: H&M is using AI to analyze customer data and predict upcoming fashion trends. The company uses this information to design products that resonate with customers and align with current trends.

3. Supply Chain Optimization: Adidas is using AI to optimize its supply chain operations by predicting demand and streamlining logistics. The company uses this technology to reduce waste, save costs, and improve efficiency.

4. Personalized Shopping Experience: Stitch Fix is using AI to provide customers with personalized styling recommendations. The company uses machine learning algorithms to analyze customer data and recommend products based on their style preferences.

Case studies of fashion companies using AI:

1. Farfetch: Farfetch is an online luxury fashion retailer that uses AI to provide customers with personalized shopping experiences. The company uses machine learning algorithms to analyze customer data and recommend products based on their style preferences.

2. Nike: Nike is using AI to design new products and improve the customer experience. The company uses machine learning algorithms to analyze customer data and design products that align with customer preferences.

Future fiction scenario of a world where AI designs and produces fashion:

In a future scenario, AI technologies could completely transform the fashion industry by automating the entire fashion design process. Customers could interact with virtual fashion designers who use machine learning algorithms to create custom designs based on their preferences. AI-powered robots could produce these designs, reducing the need for human labor and reducing waste.

Conclusion: The fashion industry is at the forefront of AI innovation, using machine learning algorithms to improve design, production, and customer engagement. AI technologies offer unprecedented opportunities to create more innovative, sustainable, and personalized fashion products. As the technology advances, we can expect to see even more exciting developments in the industry. However, the industry must ensure that the adoption of AI technologies is responsible and sustainable, considering environmental and social implications.

Chapter 14: AI and Art

Introduction: Artificial intelligence (AI) has transformed multiple industries, and the art world is no exception. AI has made it possible to explore new dimensions in art, including generative art, art restoration, and authentication. Artists and institutions are adopting AI as a tool for creative expression and enhancing their work. This chapter explores how AI is changing the art world and the implications of this technology for artists, art institutions, and society.

Explanation of how AI is being used in art: AI is being used in various ways in art. One of the most exciting ways is generative art, which involves using AI algorithms to create new works of art. These algorithms analyze existing art and use that information to create new pieces. Another use of AI in art is art restoration. Art restoration requires a great deal of skill and is time-consuming. However, AI can speed up the process by using machine learning to restore damaged artwork. Additionally, AI is being used for art authentication, where it can detect forgeries by analyzing artwork's materials, style, and other attributes.

Examples of how AI is being used in art: One example of AI being used in art is the work of

Refik Anadol. Anadol uses AI algorithms to create stunning visualizations of data, such as the movement of people in public spaces or the sound waves of a symphony. He feeds the data into the algorithm, which then generates a unique work of art. Another example is the Metropolitan Museum of Art, which is using AI to analyze the colors and composition of paintings to better understand the techniques used by artists in the past.

Case studies of artists and art institutions using AI: Stitch Fix is a fashion company that uses AI to personalize its shopping experience for customers. The company uses machine learning algorithms to analyze customers' data and recommend clothes that match their preferences. Farfetch is another company that uses AI to enhance the fashion industry. The company uses machine learning to analyze fashion trends and recommend outfits for customers.

Future fiction scenario of a world where AI creates and appreciates art: In the future, it is possible that AI will become a primary source of art creation, and people will appreciate AI-generated art. AI could even become the judge of the quality of art. However, this raises ethical questions about whether AI-generated art can have the same cultural significance as art created by human artists.

Conclusion: AI is changing the art world by enabling artists and institutions to explore new dimensions in art. AI algorithms are being used to create new works of art, restore damaged artwork, and authenticate art. However, the implications of AI for the art world and society are still being explored. As AI technology continues to develop, it is essential to consider its impact on art and the cultural significance of AI-generated art.

Chapter 15: AI and Journalism and Media

Introduction: Artificial intelligence (AI) is being increasingly used in journalism and media to improve the production, delivery, and consumption of news and other media content. AI is being used to generate news articles, fact-check claims, provide personalized news and media recommendations, and analyze and research content. As with other industries, the use of AI in journalism and media is expected to increase in the coming years.

Explanation of how AI is being used in journalism and media: AI is being used in several ways in journalism and media. One of the most common uses of AI in journalism and media is to generate news articles. AI algorithms can analyze data from various sources, including social media and news feeds, and create articles that are readable, informative, and objective. AI can also be used to fact-check news stories and claims, providing an objective analysis of the accuracy of information.

Another use of AI in journalism and media is in content creation. AI algorithms can analyze data on user preferences and produce media content that is personalized to the user's interests. This can include articles, videos, and podcasts, among others. AI can also be used to analyze and research content, providing insights into the impact of media on society.

Examples of how AI is being used in journalism and media: Associated Press (AP) has been using AI to generate news articles for several years. The AP's AI algorithm, known as Automated Insights, analyzes data from various sources and produces news articles that are published on the wire service's website. The algorithm is capable of generating thousands of articles each month, covering topics ranging from earnings reports to sports.

The Washington Post has also been using AI to generate news articles. The Post's AI algorithm, known as Heliograf, was used during the 2016 U.S. presidential election to provide real-time coverage of the election results. The algorithm produced articles that were published on the Post's website and social media channels.

CNN has been using AI to personalize its news and media recommendations. The news network's AI algorithm analyzes user behavior and preferences and provides personalized news recommendations based on those factors. CNN has also used AI to analyze its own content, providing insights into how its content is consumed and shared by users.

Case studies of news organizations and companies using AI: The New York Times has been using AI to produce audio versions of its articles. The newspaper's AI algorithm, known as Athena, analyzes articles and produces audio recordings that are published on the Times' website and mobile app.

Another news organization using AI is the BBC. The British broadcaster has been using AI to produce video clips for its news stories. The BBC's AI algorithm, known as Juicer, analyzes audio and video content and selects the most relevant clips to include in news stories.

Future fiction scenario of a world where AI is the journalist: In a future world, AI may become the primary source of journalism. AI algorithms could be programmed to produce news articles, videos, and other media content on a constant basis, covering all areas of interest to the public. AI could also be used to analyze and fact-check news stories, ensuring that all information is accurate and objective.

Conclusion: AI is becoming an increasingly important tool in journalism and media. AI algorithms can generate news articles, fact-check claims, personalize news and media recommendations, and analyze and research content. As AI technology continues to advance, its use in journalism and media is expected to increase.

Chapter 16: AI and Transportation

Introduction: Artificial intelligence (AI) is rapidly transforming the transportation industry, with the potential to improve safety, reduce congestion, and enhance efficiency. The use of AI is revolutionizing the way we move people and goods around the world, from self-driving cars to smart traffic management systems. In this chapter, we will explore the various ways in which AI is being used in the transportation industry, as well as the benefits and challenges associated with this technology.

Explanation of how AI is being used in the transportation industry: AI is being used in transportation in a variety of ways. One of the most significant applications is in autonomous vehicles, where AI is used to enable cars to operate without human intervention. Self-driving cars rely on a variety of sensors, cameras, and other technologies to navigate roads and make decisions based on real-time data. Additionally, AI is being used to improve traffic management systems, optimize routes, and reduce congestion. AI can analyze data on traffic patterns, weather, and other factors to make real-time adjustments to traffic lights, speed limits, and other factors. Another area where AI is being used in transportation is predictive maintenance. AI can be used to monitor and analyze data from vehicles to predict when maintenance is needed, helping to reduce downtime and increase efficiency.

Examples of how AI is being used in the transportation industry: There are numerous examples of how AI is being used in transportation. For instance, Tesla has been at the forefront of self-driving car technology, with its Autopilot feature enabling semi-autonomous driving on highways. Google's Waymo has also been working on autonomous vehicle technology, having launched its self-driving car service in Phoenix, Arizona. AI is also being used in public transportation systems, such as the Singapore Land Transport Authority's use of AI for real-time traffic monitoring and predicting congestion. Predictive maintenance is another area where AI is being used, with companies such as GE Aviation using AI to analyze data from airplane engines to predict when maintenance is needed.

Case studies of transportation companies and governments using AI: One example of a transportation company using AI is Daimler AG, the parent company of Mercedes-Benz. Daimler has been working on self-driving truck technology, which it plans to deploy in the near future. The company has also been using AI to optimize its supply chain and reduce costs. Another example is the Singapore Land Transport Authority, which has been using AI for real-time traffic monitoring and predicting congestion. The system uses data from various sources, including GPS, cameras, and sensors, to provide real-time traffic updates to drivers and help reduce congestion.

Future fiction scenario of a world where AI controls transportation systems: In a future where AI controls transportation systems, the benefits could be significant. AI could optimize traffic flow, reduce congestion, and improve safety by eliminating human error. However, there are also concerns about the implications of a world where AI controls transportation systems. For instance, there could be privacy concerns associated with the collection of data by autonomous vehicles. Additionally, there could be ethical considerations surrounding the use of AI in transportation, such as decisions about who is responsible in the event of accidents involving self-driving cars.

Conclusion: AI is rapidly transforming the transportation industry, with the potential to revolutionize the way we move people and goods around the world. While there are significant benefits associated with the use of AI in transportation, there are also challenges that need to be addressed. As the technology continues to evolve, it will be important to carefully consider the implications of AI for the transportation industry, and to ensure that these technologies are used in ways that are safe, ethical, and beneficial for society as a whole.

Chapter 17: AI and Human Augmentation

Introduction: Artificial intelligence (AI) is revolutionizing the way we think about human augmentation, providing a range of exciting possibilities for enhancing human capabilities. AI-powered devices and technologies can augment and support human cognitive and physical abilities, improving people's lives in many ways. This chapter will explore the current state of AI and human augmentation, with examples of how AI is being used for prosthetics, brain-computer interfaces, and human enhancement. The chapter will also examine case studies of companies and organizations that are using AI for human augmentation, including Neuralink and the US Department of Defense. Finally, the chapter will conclude with a discussion of the future of AI and human augmentation, exploring the potential benefits and risks of a world where AI merges with humans.

Explanation of How AI is Being Used to Augment Human Capabilities: AI is being used to augment human capabilities in several ways. One of the most exciting areas of AI and human augmentation is in the field of prosthetics. AI-powered prosthetics can provide a more natural and responsive experience for users, allowing them to control their prosthetic limbs with their minds or even their emotions. These devices can also learn from the user's movements and adapt to their needs over time, providing a customized experience that is tailored to their unique needs.

Another area where AI is being used for human augmentation is in the development of brain-computer interfaces (BCIs). BCIs can help people with disabilities communicate and interact with the world around them more easily. They can also be used to control devices, such as prosthetic limbs, with the power of the mind. BCIs can also help researchers better understand the brain and its functions, leading to new treatments for neurological disorders.

AI is also being used for human enhancement, with the potential to improve our cognitive and physical abilities. For example, AI-powered devices can help people with memory loss, allowing them to recall important information more easily. AI can also be used to enhance athletic performance, by analyzing data on an athlete's movements and providing feedback on how to improve technique and performance. AI-powered devices can also improve physical rehabilitation, by providing personalized training and support to help people recover from injuries or illnesses.

Examples of How AI is Being Used for Prosthetics, Brain-Computer Interfaces, and Human Enhancement: One of the most exciting examples of AI and human augmentation is in the field of prosthetics. For example, the DEKA Arm System is a prosthetic arm that uses AI to provide a more natural and responsive experience for users. The device uses sensors to detect the user's movements and respond in real-time, allowing them to control the prosthetic arm with their thoughts.

Another example of AI-powered human augmentation is in the development of brain-computer interfaces. The BrainGate system is a BCI that allows people with paralysis to control devices with their minds. The system uses implanted electrodes to record brain activity and translate it into commands that can be used to control a computer or other devices.

In the field of human enhancement, one example is the development of AI-powered memory aids. The MemorEM system is a wearable device that uses AI to help people with memory loss remember important information. The device records information throughout the day, and then uses AI algorithms to suggest information that the user may need to remember at a later time.

Case Studies of Companies and Organizations Using AI for Human Augmentation: One of the most well-known companies working in the field of AI and human augmentation is Neuralink. Founded by Elon Musk, Neuralink is developing a brain-machine interface that could one day allow humans to communicate directly with computers and other devices using their thoughts. The company is also working on developing AI-powered prosthetics that can be controlled with the mind.

As AI technology advances, it has the potential to revolutionize the way we interact with the world and even ourselves. The use of AI for human augmentation is a rapidly growing field that has already shown promise in improving the lives of people with disabilities and enhancing human performance.

One of the most well-known applications of AI for human augmentation is in prosthetics. With the help of AI, prosthetic limbs can be controlled more intuitively and precisely than ever before. For example, a neural interface can be used to connect a prosthetic arm directly to the user's nervous system, allowing them to control the limb with their thoughts. This technology has the potential to dramatically improve the quality of life for amputees and people with other disabilities.

Another area where AI is being used for human augmentation is in brain-computer interfaces (BCIs). BCIs use a combination of sensors and algorithms to translate brain activity into commands that can control devices or even communicate with other people. This technology has the potential to revolutionize communication for people with conditions such as locked-in syndrome or severe paralysis.

AI is also being used for human enhancement, with the goal of improving physical and cognitive performance beyond what is normally possible. For example, athletes could use AI-powered tools to optimize their training and nutrition plans for peak performance. Similarly, workers in high-stress environments could use AI to monitor their stress levels and optimize their work patterns for better productivity and well-being.

However, the use of AI for human augmentation raises ethical concerns about issues such as privacy, consent, and equitable access to technology. As this technology becomes more widespread, it will be important to address these concerns and ensure that everyone can benefit from these advances in a responsible and ethical manner.

Case studies of companies and organizations using AI for human augmentation include Neuralink, a company founded by Elon Musk that aims to create a brain-machine interface that can link human brains directly to computers, and the US Department of Defense, which is researching ways to enhance soldiers' physical and cognitive abilities through technology.

In a future fiction scenario, AI could merge with humans, leading to a world where people have superhuman capabilities and new forms of consciousness. However, this raises complex ethical and philosophical questions about what it means to be human and the potential risks and benefits of such a scenario.

In conclusion, the use of AI for human augmentation has the potential to revolutionize the way we interact with the world and enhance human capabilities. However, it is important to address ethical concerns and ensure that everyone can benefit from these advances in a responsible and equitable manner.

Chapter 18: AI and Cybersecurity

Introduction: The emergence of artificial intelligence (AI) has significantly impacted cybersecurity, changing the way security professionals approach the problem of protecting sensitive data and infrastructure from cyber threats. AI is increasingly being used in cybersecurity to identify threats, predict vulnerabilities, and detect and respond to attacks more effectively and efficiently than human analysts. However, AI itself presents new security challenges that must be addressed, as cybercriminals can exploit vulnerabilities in AI systems to cause damage and steal data. This chapter explores the use of AI in cybersecurity, the challenges of securing AI systems, and the potential for AI to combat cybercrime.

Explanation of how AI is being used in cybersecurity: AI is used in cybersecurity to analyze large amounts of data and identify patterns and anomalies that could indicate an attack or vulnerability. Machine learning algorithms can learn from previous cyber-attacks and use that knowledge to predict and prevent future attacks. AI can also be used to automate security processes and quickly respond to threats, reducing the time it takes to detect and remediate attacks.

Explanation of cybersecurity and the challenges of securing AI systems: Cybersecurity involves protecting computer systems, networks, and data from unauthorized access, theft, and damage. AI can enhance cybersecurity by automating security processes and providing advanced threat detection and response capabilities. However, AI itself presents new security challenges. Machine learning algorithms can be manipulated by cybercriminals to cause damage or steal data. For example, attackers can exploit weaknesses in AI models to bypass security systems, evade detection, or cause false alarms.

Examples of how AI is being used for threat detection, vulnerability assessment, and network security: AI is being used for threat detection by analyzing network traffic and identifying patterns that could indicate an attack. For example, AI can be used to detect malware, phishing attacks, and other types of cyber threats. AI can also be used for vulnerability assessment by identifying weaknesses in systems that could be exploited by attackers. Additionally, AI is used for network security by analyzing network traffic and identifying suspicious behavior or anomalies.

Case studies of companies and organizations using AI for cybersecurity: Companies and organizations are increasingly using AI for cybersecurity. Darktrace is an example of a company that uses AI to detect and respond to cyber threats. Their software uses machine learning algorithms to analyze network traffic and detect anomalies that could indicate an attack. Cylance is another example of a company that uses AI for cybersecurity. They use machine learning algorithms to detect and prevent malware attacks.

The US Department of Defense is also using AI for cybersecurity. They are developing AI tools to automate the detection and response to cyber threats, as well as identify vulnerabilities in their systems.

Future fiction scenario of a cyberattack on a global AI network and AI fights cybercrime: In a future scenario, a global AI network that controls critical infrastructure and systems is attacked by a group of sophisticated cybercriminals. The attackers use AI-based malware that can evade detection and spread rapidly across the network. The AI network responds by using advanced machine learning algorithms to analyze the malware and identify its source. The AI network also uses AI-based response systems to quickly contain and remediate the attack, preventing the cybercriminals from causing any significant damage.

Conclusion: AI is transforming the cybersecurity landscape by providing advanced threat detection and response capabilities. However, AI itself presents new security challenges that must be addressed. As AI continues to evolve, it will become increasingly important to secure AI systems against cyber threats. The potential for AI to fight cybercrime is immense, and organizations must continue to invest in AI-based security solutions to stay ahead of the evolving threat landscape.

Chapter 19: AI and Entertainment

Introduction: Artificial Intelligence (AI) has the potential to transform the entertainment industry, creating new opportunities for creativity, personalization, and engagement. From music and video creation to virtual assistants, AI is revolutionizing the way we consume and interact with entertainment. In this chapter, we will explore how AI is being used in the entertainment industry and its impact on the future of entertainment.

Explanation of how AI is being used in the entertainment industry: AI is being used in the entertainment industry in a variety of ways. One of the most significant areas of impact is in music and video creation. AI is being used to create music and video content that is personalized, unique, and engaging. For example, AI algorithms can analyze music patterns and create new compositions that sound like they were composed by humans. Similarly, AI is being used in video creation to generate realistic scenes and characters.

AI is also being used to develop virtual assistants that can help consumers find the content they want to watch. These assistants use natural language processing and machine learning algorithms to understand user preferences and recommend relevant content. This technology is being used by companies like Amazon and Google to create virtual assistants that can interact with users in a personalized and engaging way.

Another area where AI is being used is in video game design. AI is being used to create more realistic and immersive game environments by generating content and characters that respond dynamically to user actions. AI is also being used to develop smarter and more challenging opponents in games.

Examples of how AI is being used in the entertainment industry: One example of how AI is being used in the entertainment industry is in music composition. A startup called Amper Music has developed an AI platform that can create personalized music compositions based on user preferences. The platform uses deep learning algorithms to analyze music patterns and generate new compositions that sound like they were composed by humans.

Another example of how AI is being used in the entertainment industry is in virtual assistants. Amazon's Alexa and Google's Assistant are examples of virtual assistants that use natural language processing and machine learning algorithms to understand user preferences and recommend relevant content. These assistants can also be used to control smart home devices, play music, and answer questions.

In the video game industry, AI is being used to create smarter opponents and more dynamic game environments. For example, the game AI in Forza Motorsport 7 uses machine learning algorithms to analyze player behavior and create more challenging opponents.

Case studies of companies and organizations using AI for entertainment: Netflix is a prime example of a company that is using AI to personalize the entertainment experience for users. The company uses machine learning algorithms to analyze user behavior and recommend movies and TV shows that are relevant to their interests. This technology has helped Netflix become one of the most popular streaming services in the world.

Amazon is another company that is using AI to personalize the entertainment experience for users. The company's Alexa virtual assistant uses natural language processing and machine learning algorithms to understand user preferences and recommend relevant content. Amazon is also using AI in its video game development studio,

Amazon Game Studios, to create more immersive and engaging games.

Future fiction scenario of a world where AI creates all our entertainment: In the future, AI could be used to create all our entertainment. Imagine a world where AI algorithms generate music, movies, TV shows, and video games that are personalized to our interests and preferences. These entertainment experiences would be created in real-time and delivered directly to our devices. AI could also be used to create virtual worlds and immersive experiences that allow us to escape reality and explore new environments.

Conclusion: AI is transforming the entertainment industry in significant ways, from music and video creation to virtual assistants and video game design. As AI technology continues to evolve, it has the potential to create new opportunities for creativity, personalization, and engagement in the entertainment industry. In addition to creating new entertainment content, AI is also being used to improve the viewing experience. For example, AI-powered recommendation systems are used to suggest movies, TV shows, and other content based on user preferences, viewing history, and other data. This technology is used by companies such as

Netflix and Amazon to help users discover new content and keep them engaged.

AI is also being used in video game design to create more realistic and immersive gaming experiences. Game developers are using AI to create better graphics, physics, and AI-controlled characters that are more lifelike and intelligent. For example, the game "AI Dungeon" uses AI to generate a unique and evolving storyline based on player input, allowing for an almost limitless number of possible outcomes.

Chapter 20: AI in Manufacturing and Supply Chain

Introduction: Artificial Intelligence (AI) has been transforming the manufacturing and supply chain industries by improving the overall productivity, efficiency, and cost-effectiveness of operations. By leveraging AI, manufacturers can optimize their supply chain processes, reduce production times, and minimize costs while ensuring the highest levels of quality control. In this chapter, we will explore how AI is being used in the manufacturing and supply chain industries, and provide examples of successful AI implementations, along with their benefits.

Explanation of how AI is being used in manufacturing and supply chain: AI is being used in manufacturing and supply chain in a variety of ways, such as predictive maintenance, quality control, and supply chain optimization. AI algorithms are able to analyze vast amounts of data in real-time and provide insights into potential issues before they occur. This allows manufacturers to take proactive measures to prevent equipment failures and minimize downtime, thereby increasing productivity and reducing costs. Additionally, AI can be used for quality control by detecting defects and anomalies in real-time, allowing manufacturers to fix issues before they become larger problems.

Supply chain optimization is another area where AI is being applied. With the help of AI, manufacturers can optimize their inventory levels, track products throughout the supply chain, and identify potential bottlenecks in the production process. This enables manufacturers to improve delivery times, reduce inventory costs, and ensure customer satisfaction.

Examples of how AI is being used for predictive maintenance, quality control, and supply chain optimization: One example of how AI is being used for predictive maintenance is in the automotive industry. Volkswagen Group is using AI to predict the likelihood of equipment failure in their factories. By analyzing data from sensors and other sources, AI algorithms can detect patterns that indicate impending equipment failure. This allows Volkswagen to schedule maintenance before a breakdown occurs, reducing downtime and improving productivity.

In the food industry, Nestle is using AI to improve the quality of their products. By analyzing images of food products, AI algorithms can detect defects that may not be visible to the human eye. This enables Nestle to identify quality issues before the products are shipped to customers, ensuring high levels of customer satisfaction.

Amazon is another company that has been using AI for supply chain optimization. By analyzing data from their warehouses and delivery routes, AI algorithms can optimize the delivery process, reducing costs and improving delivery times. This has allowed Amazon to become a leader in the e-commerce industry, with fast and reliable delivery times that keep customers coming back.

Case studies of manufacturing and supply chain companies using AI: Siemens is a multinational corporation that specializes in industrial manufacturing, energy, and healthcare. They are using AI to optimize their supply chain processes by predicting demand, identifying potential disruptions, and ensuring on-time delivery. By implementing AI, Siemens has been able to reduce inventory costs, increase productivity, and improve customer satisfaction.

Boeing is another company that has been using AI for manufacturing and supply chain optimization. They are using AI to optimize their production processes by analyzing data from sensors and other sources. By predicting potential issues and identifying bottlenecks, Boeing has been able to increase productivity and reduce production time. This has allowed them to stay competitive in the aerospace industry and meet the increasing demand for their products.

Conclusion: AI has the potential to revolutionize the manufacturing and supply chain industries by improving productivity, efficiency, and cost-effectiveness. By leveraging AI for predictive maintenance, quality control, and supply chain optimization, manufacturers can reduce downtime, improve product quality, and optimize their supply chain processes. With the continued development of AI technologies, we can expect to see even more advancements in these areas in the future.

Chapter 21: Impact of AI on Employment and Skills

Introduction: Artificial intelligence (AI) is rapidly changing the way we work, and many are concerned about its impact on employment and skills. Some believe that AI will lead to widespread job losses and the displacement of many workers, while others argue that it will create new job opportunities and improve working conditions. This chapter explores the potential impact of AI on employment and skills and examines how companies and organizations are responding to these challenges.

Potential Impact of AI on Employment and Skills: AI has the potential to automate many tasks currently performed by humans, leading to significant changes in the job market. Some estimates suggest that up to 40% of all jobs could be automated by 2030. However, the impact of AI on employment is likely to be complex and varied, with some jobs being eliminated, while others are created or transformed.

AI is already transforming many industries, including manufacturing, transportation, and healthcare, and the demand for workers with expertise in AI and related technologies is increasing rapidly. At the same time, many workers will need to reskill or upskill to remain competitive in the changing job market. Workers who perform routine, repetitive tasks are most at risk of job displacement, while those with skills in problem-solving, creativity, and emotional intelligence are likely to be in high demand.

Examples of How AI is Changing the Job Market and Required Skills: The impact of AI on the job market is already evident in several industries. In the manufacturing sector, for example, AI is being used to automate tasks such as quality control, predictive maintenance, and supply chain optimization. This has led to increased demand for workers with expertise in AI, data science, and related fields.

In the healthcare sector, AI is being used to improve patient outcomes, reduce costs, and increase efficiency. AI-powered medical devices and diagnostic tools can analyze large amounts of patient data and assist clinicians in making better-informed decisions. However, the use of AI in healthcare also raises concerns about job displacement and the impact on patient care.

In the financial sector, AI is being used to automate tasks such as fraud detection, credit scoring, and investment management. This has led to increased demand for workers with expertise in

AI, data science, and finance. However, the use of AI in finance also raises concerns about job displacement and the impact on the stability of financial markets.

Organizations Addressing the Impact of AI on Employment and Skills: Many organizations are taking steps to address the impact of AI on employment and skills. The World Economic Forum, for example, has launched a reskilling initiative aimed at helping workers develop the skills they need to remain competitive in the changing job market. The initiative focuses on providing training in areas such as AI, data science, and cybersecurity.

IBM has also launched a reskilling initiative aimed at helping workers develop the skills they need to succeed in the era of AI. The initiative offers online training in areas such as AI, cloud computing, and blockchain, and is designed to help workers transition to new roles or industries.

Conclusion: AI is already transforming many industries and is likely to have a significant impact on employment and skills. While some jobs may be eliminated or transformed, many new job opportunities will also be created. However, workers will need to reskill or upskill to remain competitive in the changing job market.

Organizations, governments, and individuals will need to work together to address the challenges posed by AI and ensure that workers have the skills they need to succeed in the era of AI.

Chapter 22: AI in Science and Technology Research

Introduction: Artificial intelligence (AI) has rapidly transformed the scientific and technological research landscape, making research more efficient, effective, and accurate. AI is becoming increasingly relevant in areas such as physics, chemistry, biology, engineering, and many other fields. This chapter will discuss how AI is being used in scientific research, lab and technological innovation, and the implications of this integration.

Explanation of how AI is being used in scientific research, lab, and technological innovation: One of the significant benefits of AI in scientific research is the ability to process vast amounts of data and identify patterns that may be otherwise difficult to detect manually. AI algorithms are used to analyze complex datasets, to recognize complex patterns, and to improve the accuracy of predictions. For instance, in the field of genomics, AI is being used to analyze complex data sets and predict the onset of diseases such as cancer. AI is also being used to analyze and interpret the output of experiments and simulations, such as the Large Hadron Collider experiments.

AI has also significantly improved the process of drug discovery by analyzing molecular structures, predicting drug activity, and identifying new drug targets. AI is also used to model and simulate the behavior of complex systems, such as climate change, to understand the impact of various factors and identify potential solutions. Moreover, AI is being used for automating lab processes, such as identifying compounds and running experiments, making scientific research more efficient and cost-effective.

Examples of how AI is being used for data analysis, modeling, and simulation in fields such as physics, chemistry, and engineering: In the field of physics, AI is being used to analyze data from particle accelerators, such as the Large Hadron Collider, to detect signals and interpret data. AI algorithms are also used to predict the behavior of complex systems, such as fusion reactors.

In chemistry, AI is being used to design and discover new materials, analyze molecular structures, and predict chemical reactions. AI algorithms are also used to optimize chemical reactions, reducing the time and cost of discovering new compounds.

In engineering, AI is being used to design and optimize complex systems, such as autonomous vehicles, airplanes, and energy grids. AI algorithms are used to analyze data from sensors and predict the behavior of systems in real-time, improving efficiency and reducing the risk of failures.

Examples of how AI is being used for automation, optimization, and analysis in labs and research facilities: AI is also being used for automation, optimization, and analysis in labs and research facilities. Robots equipped with AI are being used to carry out experiments, collect data, and run simulations, making lab processes more efficient and accurate. AI algorithms are also being used to optimize experiments, reducing the time and cost of research. Moreover, AI is being used to analyze images and videos, making it easier to identify patterns and structures, such as in medical imaging.

Case studies of research institutions and companies using AI in science and technology: CERN, the European Organization for Nuclear Research, uses AI to analyze data from particle accelerators, such as the Large Hadron Collider, to detect signals and interpret data. AI algorithms are also used to predict the behavior of complex systems, such as fusion reactors.

IBM Research is using AI to develop new materials and optimize chemical reactions. IBM is also developing AI algorithms to analyze and interpret the output of experiments and simulations, making research more efficient and accurate.

The Joint Center for Energy Storage Research (JCESR) is using AI to design and optimize energy storage systems. JCESR is using AI algorithms to analyze data from experiments and simulations, predict the behavior of systems, and design new materials.

The Broad Institute is using AI to analyze genomic data, predict the onset of diseases, and identify new drug targets. The institute is also using AI to develop new drugs and optimize drug discovery processes.

Future fictions: In the future, AI is expected to play an increasingly important role in scientific and technological research. It is possible that in the future, AI could become a driving force behind the development of new technologies and the advancement of scientific knowledge.

One possible future fiction scenario is that AI could be used to develop new materials and substances that are stronger, lighter, and more durable than existing materials. For example, AI could be used to develop new materials for use in aerospace, automotive, and construction industries. These materials could be designed to be more efficient and cost-effective, reducing energy consumption and environmental impact.

Another possible scenario is that AI could be used to design and develop new drugs and medical treatments. AI could be used to analyze large datasets of genetic information and medical records, identifying patterns and potential treatments that may have been missed by human researchers. This could lead to the development of personalized medicine, tailored to an individual's unique genetic makeup and medical history.

Finally, AI could be used to develop new renewable energy technologies. AI could be used to analyze weather patterns and renewable energy sources such as solar, wind, and geothermal energy, optimizing the production and distribution of renewable energy. This could help reduce reliance on fossil fuels and mitigate the effects of climate change.

While these scenarios may seem like science fiction, they are within the realm of possibility given the rapid advancements being made in AI research and development. As AI becomes more sophisticated and powerful, it has the potential to revolutionize the way we approach scientific and technological research.

Chapter 23: AI in War and Defense

Introduction: The use of artificial intelligence (AI) in war and defense has been increasing rapidly in recent years. AI is being used in various military applications such as autonomous systems, decision-making, and intelligence analysis. The adoption of AI in defense has the potential to transform warfare, providing new capabilities and changing the nature of conflicts. This chapter provides an overview of how AI is being used in military applications, along with examples and case studies.

Explanation of how AI is being used in military applications: AI is being used in various military applications to augment human decision-making and enhance military capabilities. One of the key areas of AI application is in autonomous systems, which includes drones, ground vehicles, and unmanned submarines. These systems are equipped with AI algorithms that enable them to operate without human intervention, and perform various tasks such as reconnaissance, surveillance, and attack. The use of autonomous systems reduces the risk to human life and provides the military with the ability to perform missions that are otherwise impossible or too dangerous for humans.

Another area where AI is being used in military applications is in decision-making. AI algorithms can analyze large amounts of data and provide recommendations for military commanders. These recommendations can be used for planning operations, evaluating the effectiveness of tactics, and identifying potential threats. AI can also be used to simulate scenarios and predict the outcomes of military actions, allowing for more informed decision-making.

AI is also being used in intelligence analysis to process and analyze large amounts of data collected from various sources, such as sensors, drones, and satellites. The analysis of this data can provide valuable insights into the activities of enemy forces, and enable military commanders to make informed decisions about operations.

Examples of how AI is being used for autonomous systems, decision-making, and intelligence analysis: The use of AI in military applications is not limited to one country or region. Several countries are actively developing and deploying AI-based military systems. For example, the United States military is using drones equipped with AI algorithms for reconnaissance and surveillance missions. These drones are capable of autonomously identifying and tracking targets, providing real-time intelligence to military commanders.

China is also investing heavily in AI-based military systems, including autonomous submarines and drones. These systems are being developed to perform various tasks such as surveillance, reconnaissance, and attack.

Israel is using AI algorithms to analyze intelligence data collected from various sources, including drones and satellites. The analysis of this data enables the military to identify potential threats and take preemptive action.

Case studies of military organizations and companies using AI: The United States Department of Defense (DoD) is actively investing in AI-based military systems. The DoD has established the Joint Artificial Intelligence Center (JAIC) to develop and deploy AI technologies for military applications. The JAIC is working on various projects, including autonomous systems, decision-making, and intelligence analysis.

Lockheed Martin, a leading defense contractor, is also investing in AI-based military systems. The company is developing autonomous systems such as drones and ground vehicles, along with AI algorithms for decision-making and intelligence analysis.

Conclusion: The use of AI in military applications has the potential to transform warfare, providing new capabilities and changing the nature of conflicts. However, the adoption of AI in defense also raises ethical and legal concerns. The development and deployment of AI-based military systems should be guided by principles of ethical use, transparency, and accountability. The military organizations and companies developing AI-based military systems must work with governments and international organizations to establish guidelines and regulations for their use.

Chapter 24: AI in Space Exploration

Introduction: Artificial Intelligence (AI) is playing a crucial role in space exploration, revolutionizing how we explore and understand the universe beyond our planet. The application of AI in space missions has enhanced the accuracy and efficiency of data analysis, decision-making, and control of autonomous systems. This chapter explores the current and potential use of AI in space exploration, highlighting its benefits and challenges.

Explanation of how AI is being used in space exploration:

AI is being used in several aspects of space exploration, including mission planning, spacecraft control, data analysis, and autonomous systems. For instance, AI algorithms are used to process and analyze the vast amounts of data gathered from space, identifying patterns and anomalies that might not be immediately recognizable to humans. Moreover, AI is used to automate tasks such as spacecraft navigation, fuel management, and propulsion systems, reducing the workload of human operators and improving mission efficiency.

Examples of how AI is being used for mission planning, autonomous systems, and data analysis in space exploration:

One of the most significant applications of AI in space exploration is in autonomous systems. AI algorithms can control spacecraft and rovers, enabling them to navigate harsh and remote environments without human intervention. For example, NASA's Mars rovers, Curiosity and Perseverance, use AI to make decisions about where to move and what to investigate based on images and other data collected during their missions.

AI is also being used for mission planning, assisting engineers and scientists in designing optimal mission trajectories and schedules. NASA's Deep Space Network, which is responsible for communicating with spacecraft throughout the solar system, uses AI to optimize data transfer rates and minimize signal delays.

Moreover, AI is being used to process and analyze the vast amounts of data collected during space missions. For example, the European Space Agency's Gaia satellite, which is mapping the Milky Way galaxy, generates approximately 100 GB of data per day. AI algorithms are used to identify patterns and structures in the data, enabling astronomers to make new discoveries about the structure and evolution of our galaxy.

Case studies of space exploration organizations and companies using AI: NASA has been a leader in using AI for space exploration. For instance, NASA's Earth Observing System Data and Information System (EOSDIS) uses AI to automate the processing and analysis of Earth observation data, allowing scientists to identify changes in the environment, such as deforestation or ocean temperature changes. Additionally, NASA's Autonomous Sciencecraft Experiment (ASE) is using AI to enable spacecraft to conduct scientific investigations without human intervention.

SpaceX is another company that has been using AI for space exploration. The company's autonomous droneships use AI algorithms to navigate and land rocket boosters at sea, a crucial aspect of SpaceX's reusable rocket technology.

Future fiction scenario of space exploration using AI: In the future, AI could revolutionize space exploration, allowing for more ambitious and complex missions. For instance, AI could be used to enable fully autonomous spacecraft that can repair themselves and make decisions about their missions without human intervention.

Moreover, AI could enable more efficient and accurate search for extraterrestrial life, using machine learning algorithms to identify signs of life in vast amounts of data collected during space missions.

Conclusion: The use of AI in space exploration is transforming how we explore and understand the universe. AI algorithms are being used to automate tasks, improve mission efficiency, and enhance the accuracy and efficiency of data analysis. As AI technology continues to evolve, it will likely play an even more significant role in space exploration, enabling more ambitious and complex missions, and unlocking new discoveries about our universe. However, AI also poses challenges, such as ensuring the safety and reliability of autonomous systems and addressing ethical concerns related to the use of AI in space exploration.

Chapter 25: The Future of AI

Introduction: Artificial intelligence (AI) has come a long way since its inception, and its future is promising, but also uncertain. The continued development of AI has been driven by technological advancements, the abundance of data, and the increasing demand for AI in various industries. This chapter will explore the potential future innovation and developments in AI, the possibilities of AI surpassing human intelligence and the concept of a singularity, and the potential implications of advanced AI for society.

Future Innovation and Developments in AI: The future of AI is exciting and promising, with several areas of potential innovation and developments. One area of AI innovation is the development of more advanced and sophisticated algorithms that can learn and adapt faster and more accurately than ever before. Another area of innovation is the integration of AI with other emerging technologies such as blockchain, quantum computing, and the Internet of Things (IoT) to create more powerful and intelligent systems. Additionally, there will be a continued emphasis on ethical considerations, including privacy and data protection, as well as transparency in AI decision-making.

AI Surpassing Human Intelligence: The idea of AI surpassing human intelligence has been a topic of much debate and speculation. Many experts believe that it is inevitable that AI will surpass human intelligence at some point in the future. Some believe that this could happen within the next few decades, while others predict it could take much longer. The concept of a singularity, where AI becomes self-improving and capable of recursive self-improvement, has been proposed as a potential outcome of AI surpassing human intelligence.

Implications of Advanced AI for Society: Advanced AI has the potential to transform society in numerous ways, both positively and negatively. On the positive side, advanced AI could lead to significant advancements in healthcare, transportation, education, and other areas of human development. On the negative side, there are concerns about job displacement, inequality, and the potential misuse of AI for malicious purposes, such as cyberattacks or autonomous weapons.

Potential Future Applications of AI: There are several potential future applications of AI that could have significant impacts on society. One potential application is the development of brain-computer interfaces that could enable humans to interact with computers and machines more seamlessly. Another potential application is the development of autonomous systems, such as self-driving cars, drones, and robots that can perform complex tasks with minimal human intervention.

Risks and Benefits of Advanced AI: Advanced AI has the potential to bring significant benefits to society, but it also poses significant risks. The benefits include increased efficiency, accuracy, and productivity in various industries, while the risks include job displacement, privacy concerns, and the potential for misuse. It is important to weigh these risks and benefits carefully as AI continues to evolve and become more advanced.

Case Studies of Researchers and Organizations Exploring the Future of AI: Several researchers and organizations are actively exploring the future of AI. Elon Musk, for example, has been a vocal advocate of regulating AI to prevent it from becoming a threat to humanity. The Future of Life Institute is another organization that is focused on ensuring that the development of AI is done in a safe and ethical manner.

Future Fiction Scenario of a World where Advanced AI Shapes and AI has Surpassed Human Intelligence: In a future world where AI has surpassed human intelligence, it is difficult to predict what the implications would be. However, one possible scenario is that AI would become the dominant force in society, controlling most aspects of human life. Humans may become reliant on AI for decision-making and problem-solving, leading to a loss of autonomy and agency. Alternatively, humans could collaborate with AI in a symbiotic relationship, using AI to enhance their own abilities and to solve complex problems.

Conclusion: The future of AI is exciting, but it is also uncertain. As AI continues to evolve and become more advanced, it is important to carefully consider the potential benefits and risks that come with its development and deployment. AI has the potential to revolutionize many aspects of society, including healthcare, transportation, and education, but it also poses significant risks, such as job displacement and the potential for misuse.

It is crucial that policymakers, researchers, and industry leaders work together to develop responsible AI policies and ensure that AI is developed and deployed in an ethical and transparent manner. This includes addressing concerns such as bias, accountability, and privacy.

As AI continues to advance, it is also important to continue exploring its potential applications and to engage in ongoing discussions about the future of AI. By working together and approaching AI development and deployment with caution and responsibility, we can unlock the full potential of this technology while minimizing its risks.

Chapter 26: Conclusion: Embracing the AI Revolution

The AI revolution is well underway, and it has the potential to transform our world in unprecedented ways. Throughout this book, we have explored the impact of AI on various industries, including healthcare, finance, entertainment, and more. We have seen how AI is being used to solve complex problems, optimize processes, and enhance decision-making.

One of the most significant themes that emerged throughout this book is the need to embrace the AI revolution. While some may fear the rise of AI and its potential to replace human workers, the reality is that AI has the potential to create new job opportunities and increase productivity. It is crucial that we embrace the potential of AI and prepare for its continued growth.

At the same time, it is essential to consider the ethical implications of AI. As AI becomes more advanced, there is a risk of bias and discrimination in decision-making, as well as concerns about privacy and security. Therefore, it is crucial that we approach AI development and implementation with a responsible and ethical mindset.

In the future, we can imagine a world where humans and AI coexist harmoniously, working together to solve the world's most pressing challenges. This future may involve the development of brain-computer interfaces and other advanced technologies that allow us to communicate with machines on a more intuitive level.

Finally, we have seen numerous examples of companies and organizations leading the way in embracing the AI revolution. Google, for example, has invested heavily in AI research and development, while Amazon has implemented AI-powered robots in their warehouses to optimize their supply chain.

In **conclusion**, the AI revolution is here to stay, and it is up to us to embrace it and prepare for its continued growth. By approaching AI development and implementation with a responsible and ethical mindset, we can ensure that AI has a positive impact on our society and the world at large.

www.ingramcontent.com/pod-product-compliance
Lightning Source LLC
Chambersburg PA
CBHW071137220526
45467CB00015B/1291